中国地质大学(武汉)地质认识实习系列教材
A Series of Practice Teaching Materials for China University of Geosciences (Wuhan)

国家自然科学基金"国家基础科学人才培养基金(J1310038)"
中国地质大学(武汉)教务处"本科教学质量工程"
中国地质大学(武汉)实验室与设备管理处
联合资助

北戴河典型地质现象图册

Atlas of Typical Geological Phenomena in Beidaihe

主　编　赵俊明
Chief Editor　Zhao Junming

副主编　王家生　刘爱民　袁爱华　陈　林
Deputy Editor　Wang Jiasheng　Liu Aimin　Yuan Aihua　Chen Lin

中国地质大学出版社
CHINA UNIVERSITY OF GEOSCIENCES PRESS

图书在版编目(CIP)数据

北戴河典型地质现象图册/赵俊明主编.—武汉:中国地质大学出版社,2019.6
ISBN 978-7-5625-4573-6

Ⅰ.①北…
Ⅱ.①赵…
Ⅲ.①北戴河-地质图-图集
Ⅳ.①P562.224-64

中国版本图书馆CIP数据核字(2019)第139053号

北戴河典型地质现象图册

赵俊明 主编
王家生 刘爱民 袁爱华 陈 林 副主编

责任编辑:马 严　　责任校对:周 旭

出版发行:中国地质大学出版社(武汉市洪山区鲁磨路388号)　　邮政编码:430074
电 话:(027)67883511　　传 真:(027)67883580　　E-mail:cbb@cug.edu.cn
经 销:全国新华书店　　http://cugp.cug.edu.cn

开本:787毫米×1 092毫米 1/12　　字数:420千字 印张:13.5
版次:2019年6月第1版　　印次:2019年6月第1次印刷
印刷:武汉中远印务有限公司　　印数:1—2 200册

ISBN 978-7-5625-4573-6　　定价:138.00元

如有印装质量问题请与印刷厂联系调换

序

赵俊明教授主编的《北戴河典型地质现象图册》，是中国地质大学（武汉）北戴河地质认识实习系列教材中以精美的野外地质图片见长的一本教材，它是作者在长期伴随教师备课和学生实习过程中观察到的各类精彩地质现象和捕捉到的师生们实习生活精彩瞬间的缩影。该教材集知识性、趣味性、可读性和美育于一体，无论是对教师备课还是对同学们实习都将发挥其他教材不可替代的作用。

《北戴河典型地质现象图册》由基地概况、矿物岩石、地层和古生物、构造地质、河流地质作用、海洋地质作用、矿产资源与环境保护和旅游地质资源等章节组成，内容十分丰富。教材中各类地质照片不仅主题突出、清晰美观，而且摄影的角度和构图都颇为讲究，不少在野外难以看全、看清楚的地质现象，在本教材的照片中能得到更好的展示，简明扼要的图片文字说明，更有助于读者理解和把握照片的内容。

北戴河及邻区野外地质现象丰富，岩浆岩、沉积岩、变质岩分布广泛，从新元古代至新生代地层发育和出露良好，外动力和内动力地质作用现象典型，有着丰富多彩的海蚀地貌、多种多样的海洋生物和特征鲜明的海洋沉积以及优美的环境和宜人的气候。每年夏季北戴河及邻区都吸引了众多高校地学专业的学子来这里实习。因此，该教材的出版对我校和兄弟院校的教学实习和人才培养乃至科学普及都将发挥积极的作用。

在《北戴河典型地质现象图册》即将付印之际，谨此向读者推荐此书，并向编者表示热烈的祝贺，并希望有更好、更多的配套教材陆续面世！

2017年10月1日于武汉

Foreword

Atlas of Typical Geological Phenomena in Beidaihe, edited by Professor Zhao Junming and other authors, is characterized by awesome pictures of field geology among a Series of teaching materials for Beidaihe geological congnition practice. This atlas is an epitome of different splendid geological phenomena and wonderful moment during the field practice, captured by the authors based on a long-term accompany with teachers and students. It is an ingenious combination of being informative, interesting, readable and aesthetic. It will provide unique value for not only teaching preparation but also field practice for students.

This atlas has substantial content of the following chapters: Guide to the Base, Minerals and Rocks, Stratigraphy and Paleontology, Structural Geology, Fluvial Geological Processes, Marine Geological Processes, Mineral Resources and Environmental Protection, Geological Tourism Resources and so on. The photos in the atlas are not only attractive with prominent theme and high resolution, but also paid fastidious attention in the angle and composition of photographing. A good few of geological phenomena, which are difficult to be observed thoroughly and clearly in the field, are perfectly displayed in the atlas. Furthermore, brief and concise captions are helpful to readers for better understanding and grasping the content of photos.

There are abundant geological phenomena in Beidaihe and its adjacent areas. The magmatic rocks, sedimentary rocks and metamorphic rocks are widely distributed and well outcropped from the Neoproterozoic to the Cenozoic. Both the exogenic and endogenic geological processes are typically recorded. Every summer, Beidaihe attracts numerous college students to experience their first geological cognition practice by its rich and colorful abrasion geomorphy, a variety of marine organisms, distinct marine deposits, comfortable environments and pleasant climate. Consequently, this teaching material will play a positive role in field teaching, personnel training and science popularizing for both China University of Geosciences (Wuhan) (CUG) and other peer colleges.

On the occasion of printing of *Atlas of Typical Geological Phenomena in Beidaihe*, I strongly recommend this book to readers and express warm congratulations to the authors. And I look forward to more and better supporting teaching materials to come out in future.

Gong Yiming
October 1, 2017, Wuhan

前　言

秦皇岛市地质、地貌现象丰富，名胜古迹繁多，历史人文底蕴深厚，交通便利。中国地质大学（武汉）自1984年在北戴河建立实习基地以来，经过三十余年的野外教学实践，精选出了数十条野外地质观察路线。

针对地球科学实践性强，以及本科一年级学生野外认识实习需要大量视觉感性认知的特点，本书编者以野外地质观察路线为线索，历经数年，对实习区最主要的地质现象进行系统的拍摄，获得了大量典型地质现象图片。编者经过整理概括分类和撰写说明，编制出秦皇岛地区典型地质现象图册，为地质类专业本科生的实践教学提供系列地质现象影像图片资料，便于学生学习参考。

本书在中国地质大学（武汉）教务处进行了教学立项。本书在前人教学科研成果的基础上，在北戴河实习带队教师的协助下，由赵俊明、王家生、刘爱民、袁爱华、陈林摄制编撰完成。本书编写过程中，得到教务处杨伦处长、庞兰副处长，地球科学学院杨坤光副院长等领导的热情关怀和大力支持。桑隆康教授、龚一鸣教授审阅了全书初稿，并提出许多宝贵的意见。全国模范教师龚一鸣教授拨冗为本书撰写了序言。在此，对为本书提供帮助的所有老师和同仁们表示衷心的感谢！

编　者

2017年10月18日

Preface

Qinhuangdao is a city with abundant geological and geomorphic phenomena, deep roots of history and culture and convenient transportation. Based on the field teaching practice for more than 30 years since 1984 when the Beidaihe base was initially established by CUG, dozens of field geological excursions have been selected.

In view of the strong practicality of Earth Sciences and the requirement of visual perception for the freshmen, the authors have carried out systematic photographing according to the field geological excursions for several years and acquired a large amount of photos of typical geological phenomena in the practice area. After sorting out the photos and adding brief captions, the authors compile this atlas with photo materials of comprehensive and systematic geological phenomena for students majoring in geology.

This atlas is supported by a teaching project funded by the dean's office of CUG. Based on achievements on teaching and research by many predecessors, this atlas is compiled by Zhao Junming, Wang Jiasheng, Liu Aimin, Yuan Aihua, Chen Lin and so on with assistant of lots of teachers in Beidaihe field teaching. The preparations for this atlas were enthusiasmly concerned and greatly supported by Yang Lun, the director of dean's office of CUG, Pang Lan, the deputy Director of dean's office of CUG, Yang Kunguang, the deputy dean of School of Earth Sciences of CUG and other leaders. Professor Sang Longkang and Professor Gong Yiming reviewed the whole draft of the atlas and put forward to plenty of valuable and constructive suggestions. Professor Gong Yiming, one of the National Model Teachers, took the time to write the foreword for the atlas. Here, we express sincere thanks for all teachers and colleagues who have provided help for this atlas.

The authors

October 18, 2017

目录
CONTENT

1 基地概况

Guide to the Base

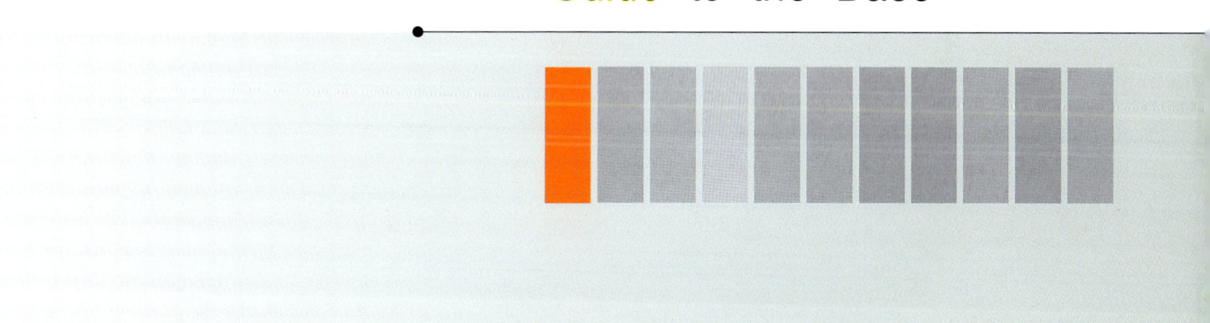

中国地质大学北戴河地质实习站位于河北省秦皇岛市海港区山东堡，东临渤海湾山东堡海滩。实习站距秦皇岛火车站15km，距北戴河火车站15km，距G1京哈高速公路入口12km。实习区内实现了公路全覆盖，交通十分便利。

Beidaihe Geological Practice Station of China University of Geosciences is located in Shandongpu, Haigang district, Qinhuangdao of Hebei Province. It faces the Shandongpu beach in Bohai Bay on the east. The Station is 15km away from Qinhuangdao railway station and the Beidaihe railway station and 12km from the entrance of Beijing-Harbin Motorway G1. The transportation is very convenient due to the full coverage of highway in the practice area.

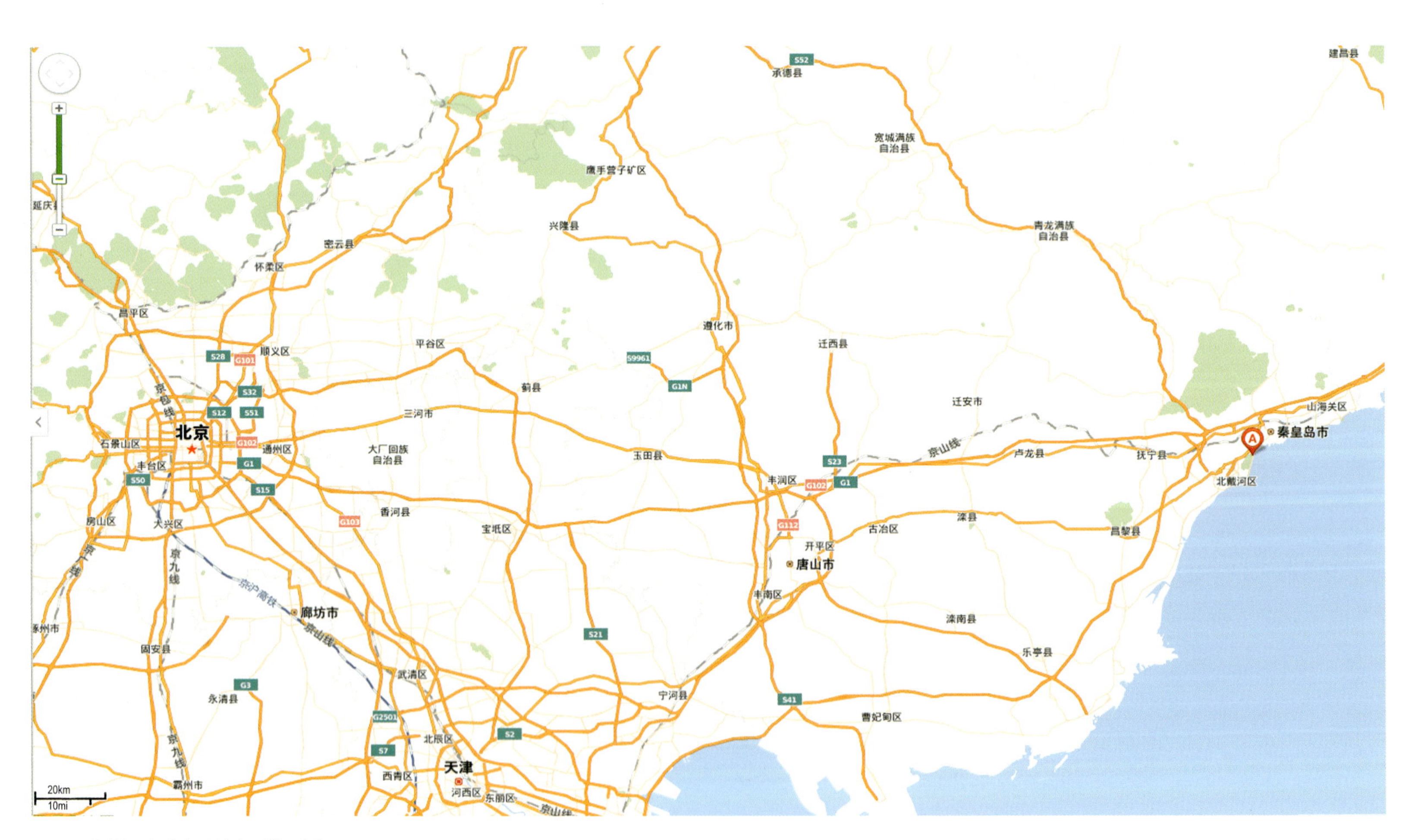

1－1　北戴河地质实习站交通位置图
Traffic location map of Beidaihe Geological Practice Station

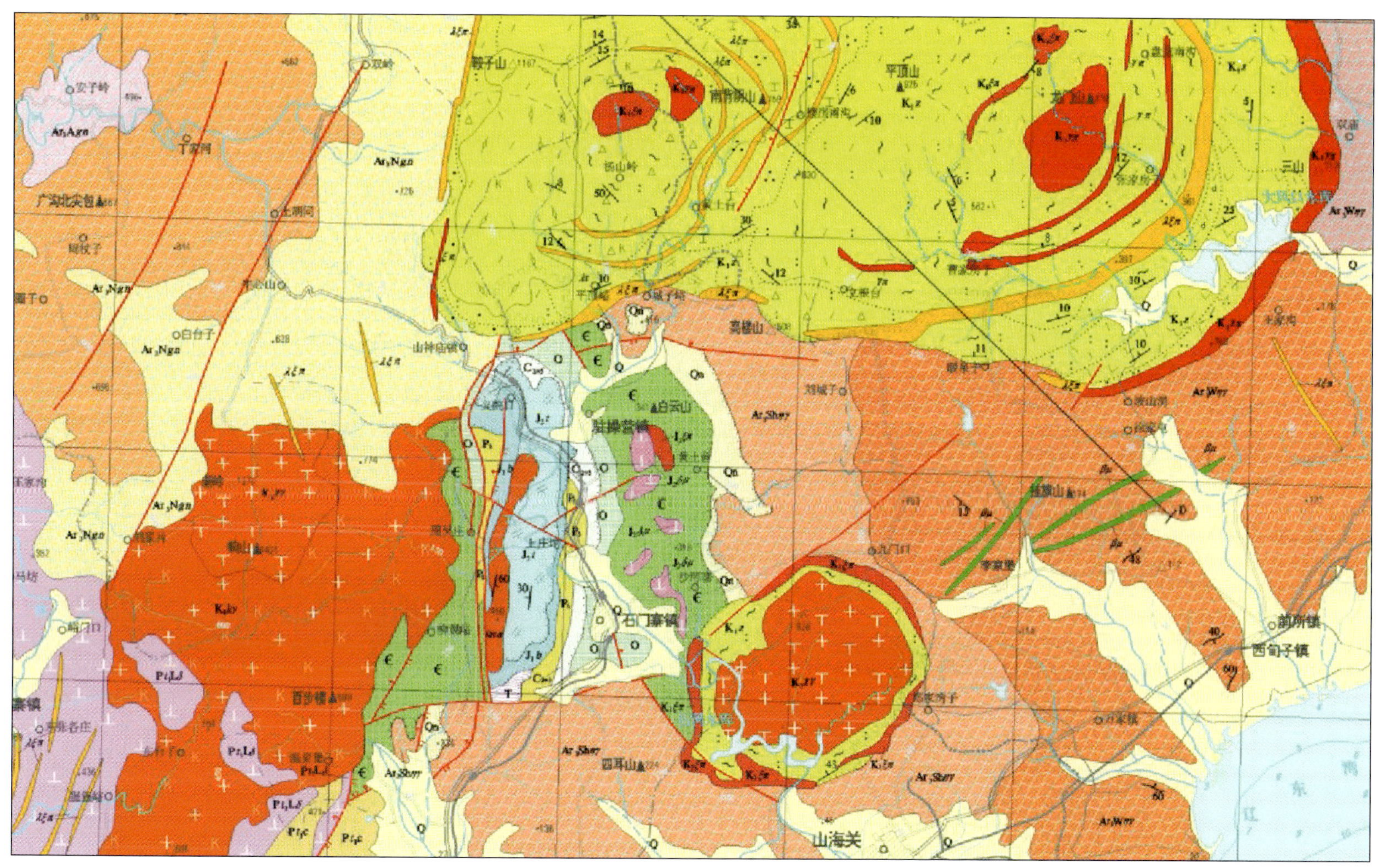

1－2　石门寨教学实习区地质简图
Geological sketch map of the Shimenzhai field teaching area

秦皇岛市地貌类型多样，山地、丘陵、平原、海岸带从北向南呈梯状分布。教学实习路线东起山海关，西至南戴河；北起柳江盆地，南至渤海海滨。东西长约35km，南北宽约25km。

实习区发育有丰富的内、外动力地质作用现象，出露了华北地区典型的新元古代以来地层，接触关系清晰。三大岩类较为齐全，其中印支－燕山期岩浆岩出露广泛。柳江向斜等构造现象清晰。海洋和河流等外动力地质作用十分发育。

There are a variety of landforms in Qinhuangdao. Mountains, hills, plains and costal zones are steppedly distributed from north to south. The field practice excursions spans a distance of 35km from the east at Shanhaiguan to the west at Nandaihe and 25km from the north of Liujiang Basin to the south along the Bohai seaside.

In the practice area, various exogenic and endogenic geological processes are developed. Typical strata of North China since the Neoproterozoic are well exposed with clear contact relationships. All the three main types of rocks can be observed including the widely distributed magmatic rocks formed during the Indosinian-Yanshanian. Structural phenomena such as Liujiang Syncline are clearly displayed. Marine and Fluvial geological processes are well developed.

1-3 北戴河地质实习站(左下)与秦皇岛市山东堡海滨相邻(2011年摄)
The Beidaihe Geological Practice Station (lower left of the photo) and its adjacent Shandongpu beach, Qinhuangdao (photographed in 2011)

1－4　建站初期的北戴河地质实习站(据1985年实习学生素描)

The Beidaihe Geological Practice Station in the early days of establishment (from the sketch by a student during field practice in 1985)

1－5　北戴河地质实习站大门(2017年摄)
The gate of the Beidaihe Geological Practice Station (photographed in 2017)

1－6　教学楼
The teaching building

1－7　宿舍
The dormitory

1－8　陈列室
The showroom

1－9　实习站大院及运动场
The courtyard and playground

1－10　花园
The garden

1－11　篮球场
The basketball court

1－12　食堂
The canteen

1－13　清晨实习的师生从这里出发
In the early morning, teachers and students gather here for setting out on field excursions

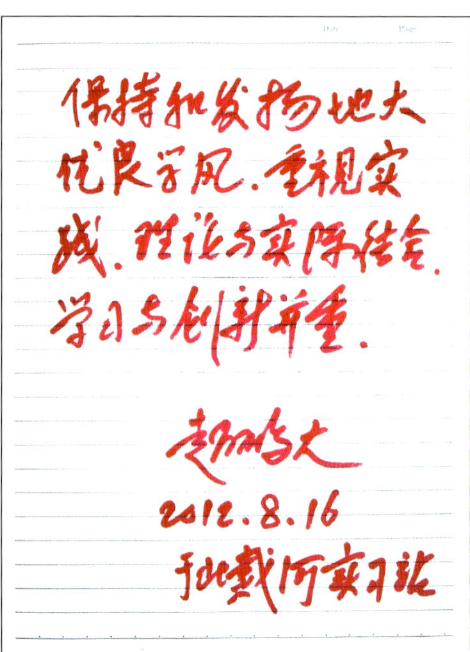

保持和发扬地大优良学风.重视实践.理论与实际结合.学习与创新并重.

赵鹏大

2012.8.16

于北戴河实习站

1－14　赵鹏大院士来实习站考察并题词
Academician Zhao Pengda came to the station and dedicated the inscription

1－15　翟裕生院士来实习站作报告
Academician Zhai Yusheng gave a presentation in the station

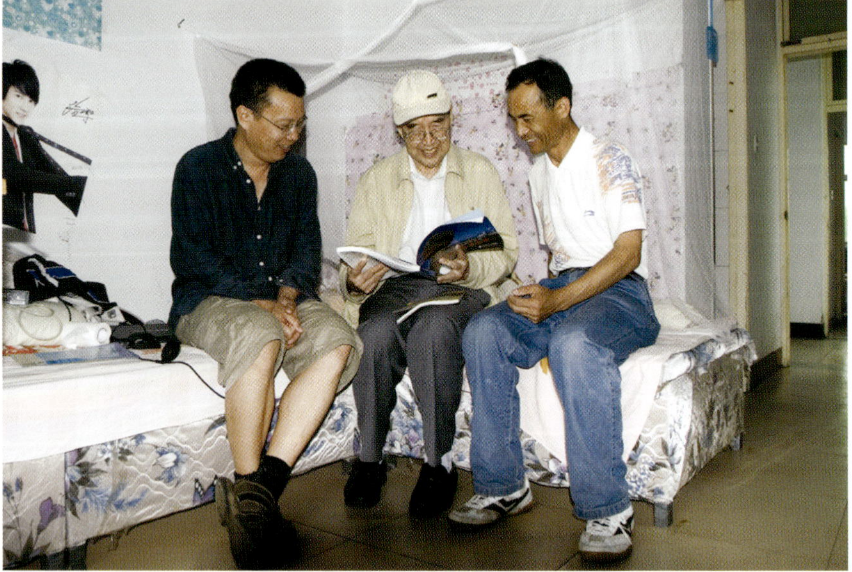

1－16　翟裕生院士来实习站考察
Academician Zhai Yusheng visited the station

1 17 教师在室内备课
The teachers were preparing for the lessons indoors

1－18 周修高教授、王红梅教授在制作实习生物标本
Professor Zhou Xiugao and Professor Wang Hongmei were making the biological specimens

1－19 教师在海滨备课
The teachers were preparing for the lessons at the seaside

1－20 刘本培教授给学生作学术报告
Professor Liu Benpei was giving an academic report for students

1－21　实习站领导作实习动员
The leader of the station was at the mobilization meeting held at the beginning of the field practice

1－22　教师带领学生观察海滨地质现象
The teachers instructed the students in observing the geological phenomena at the seaside

1－23　童金南教授给学生作现场教学
Professor Tong Jinnan was in the field teaching

1－24　龚一鸣教授给学生作现场教学
Professor Gong Yiming was in the field teaching

1－25　英国学者梅森博士在实习区野外备课
The UK researcher, Dr. Mason was preparing for the lessons in the field

1－26　留学生在野外实习
International students were in the field

1－27　学生在实习途中野外用餐
The students shared and enjoyed lunch during the field excursion

1－28　室内考试
Indoor test

1－29　教师带领学生在野外观察
The teachers instructed the students in observing in the field

2 矿物岩石

Minerals and Rocks

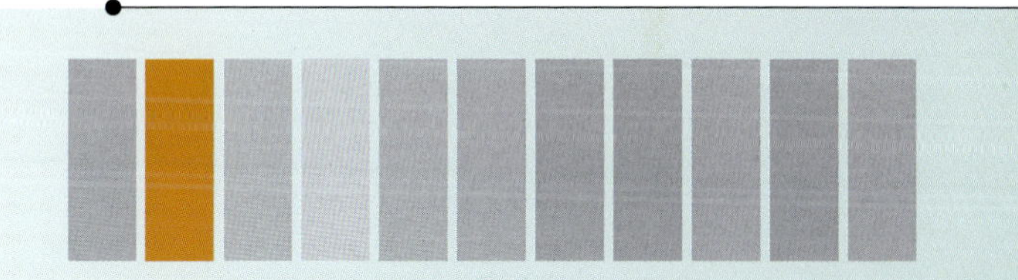

秦皇岛地区处于燕山造山带东段，东与太平洋板块相邻。活跃的内力地质作用使得岩浆岩和变质岩发育广泛，露头良好。实习区部分侵入岩结晶较充分，野外露头可以识别多种矿物。

Qinhuangdao area is situated at the east section of the Yanshan Orogenic belt and adjacent to the Pacific Plate to the east. The active endogenic geological processes brought up widely distributed and well exposed magmatic and metamorphic rocks. A part of intrusive rocks were rather completely crystallized so that it is possible to distinguish different kinds of minerals in the field outcrops.

2－1　石英，白色，块状或粒状，贝壳状断口油脂光泽（摄于北戴河骆驼石）
Quartz, a massive or granular mineral with white color and conchoidal fracture on which greasy luster can be observed (photographed at Luotuoshi in Beidaihe)

2－2　石英晶簇，单晶呈六方柱状（摄于北戴河骆驼石）
Quartz druses composed of hexagonal columnar single crystals (photographed at Luotuoshi in Beidaihe)

2－3　钾长石伟晶，肉红色，具两组解理（摄于北戴河海滨）
K-feldspar pegmatite with flesh pink color and two directions of cleavage (photographed at the Beidaihe seashore)

2－4　钾长石斑晶，肉红色，具卡氏双晶（摄于山海关燕塞湖）
K-feldspar phenocrysts with flesh pink color and displaying Carlsbad twinning (photographed at Yansaihu, Shanhaiguan)

2－5　斜长石，伟晶，灰白色(摄于北戴河海滨)
Gray and giant-crystallized plagioclase (photographed at the Beidaihe seashore)

2－6　黑云母，黑褐色片状(摄于北戴河骆驼石)
Platy dark brown biotite (photographed at Luotuoshi in Beidaihe)

2－7　角闪石，墨绿色，长柱状（摄于石门寨上庄坨）
Long columnar dark green hornblende (photographed at Shangzhuangtuo Village, Shimenzhai Town)

2－8　角闪石斑晶（摄于石门寨上庄坨）
Hornblende phenocrysts (photographed at Shangzhuangtuo Village, Shimenzhai Town)

2－9　电气石，黑色，柱状（摄于北戴河骆驼石）
Columnar black tourmaline (photographed at Luotuoshi in Beidaihe)

2－10　锆石（摄于北戴河骆驼石）
Zircon (photographed at Luotuoshi in Beidaihe)

2-11 绿泥石，黄绿色半透明硅酸盐矿物（摄于山海关燕塞湖）
Chlorite, a translucent silicate mineral with yellow-green color (photographed at Yansaihu, Shanhaiguan)

2-12 花岗岩，中粗粒结构，块状构造，主要矿物有长石、石英、黑云母等，深成酸性岩浆岩（摄于石门寨沙锅店）
Granite, a plutonic felsic magmatic rock with medium-coarse grained texture and massive structure, mainly composed of feldspar, quartz, biotite, etc (photographed at Shaguodian Village, Shimenzhai Town)

2-14 花岗岩，细粒结构，块状构造，主要矿物有长石、石英、黑云母等，深成酸性岩浆岩（摄于山海关燕塞湖）
Granite, a plutonic felsic magmatic rock with fine grained texture and massive structure, mainly composed of feldspar, quartz, biotite, etc (photographed at Yansaihu, Shanhaiguan)

2-13 海绿石，灰绿色粒状，层状硅酸盐矿物，典型海相指相矿物（摄于鸡冠山）
Glauconite, a layer silicate mineral, greyish-green, granular, a kind of typical facies mineral of marine environment (photographed at Jiguanshan)

2-15 花岗岩，细粒（摄于山海关燕塞湖）
Fine grained granite (photographed at Yansaihu, Shanhaiguan)

2－16　侵入接触关系。花岗斑岩呈岩墙侵入于奥陶系亮甲山组灰岩中，岩墙为浅肉红色斑状结构，块状构造，斑晶为钾长石和石英，基质为隐晶质（摄于石门寨沙锅店）
Intrusive contact. Granite porphyry intruding into the limestone of the Liangjiashan Formation of the Ordovician as a dyke. The granite porphyry dyke is flesh pink with porphyritic texture and massive structure. The phenocrysts are mainly K-feldspar and quartz and the groundmass is cryptocrystalline（photographed at Shaguodian Village，Shimenzhai Town）

2－17　花岗斑岩脉，侵入于亮甲山组灰岩中（摄于石门寨沙锅店）
Granite porphyry vein intruding into the limestone of the Liangjiashan Formation of the Ordovician（photographed at Shaguodian Village，Shimenzhai Town）

2－18　花岗斑岩中的石英斑晶，粒径0.1~1.2cm（摄于石门寨沙锅店）
The quartz phenocrysts in the granite porphyry with the grain size of 0.1~1.2cm（photographed at Shaguodian Village，Shimenzhai Town）

2－19　正长斑岩脉，脉状侵入于斑状正长岩中，岩脉宽约3~5cm（摄于山海关燕塞湖）
Orthophyre veins intruding into the porphyry syenite, with the width of 3~5cm (photographed at Yansaihu, Shanhaiguan)

2－20　正长斑岩，钾长石斑晶，斑状结构，块状构造，斑晶为肉红色正长石，基质为浅灰色、隐晶质（摄于山海关燕塞湖）

Orthophyre with porphyritic texture and massive structure. K-feldspar phenocrysts are mainly freshly red orthoclase and the groundmass is greyish and cryptocrystalline（photographed at Yansaihu, Shanhaiguan）

2－21　烘烤边和冷凝边，正长斑岩侵入于斑状正长岩的边界。根据其侵入接触关系，可以确定正长斑岩形成时间晚于斑状正长岩（摄于山海关燕塞湖）

Baked and chilled borders, the evidence of the Orthophyre intrusion into the porphyry syenite and of their formation order, i.e. Orthophyre was formed later than the porphyry syenite（photographed at Yansaihu, Shanhaiguan）

2－22　辉绿岩岩床和岩墙。辉绿岩，深灰色，辉绿结构，主要矿物为斜长石和辉石，侵入于早奥陶世亮甲山组灰岩中（摄于石门寨亮甲山）
Dolerite sill and dike intruding into the Early Ordovician limestone of the Liangjiashan Formation. Dolerite, a dark grey magmatic rock with diabasic texture, mainly includes plagioclase and augite (photographed at Liangjiashan, Shimenzhai Town)

2－23　辉绿岩岩脉和岩墙，侵入于亮甲山组灰岩中(摄于石门寨亮甲山)
Dolerite veins and dike intruding into the limestone of the Liangjiashan Formation(photographed at Liangjiashan, Shimenzhai Town)

2－24　侵入接触关系，辉绿岩岩墙(上)侵入于亮甲山组灰岩(下)(摄于石门寨亮甲山)
Intrusive contact between the dolerite dike (upper) and the limestone of the Liangjiashan Formation (below) (photographed at Liangjiashan, Shimenzhai Town)

2－25　细粒花岗岩(细粒0.2~2cm)，主要成分：长石、石英、黑云母(摄于山海关燕塞湖)
Fine grained granite with the grain-size of 0.2~2cm and main minerals of feldspar, quartz and biotite (photographed at Yansaihu, Shanhaiguan)

2－26　晶洞细晶花岗岩。晶洞大小为1~15mm(摄于山海关燕塞湖)
Aplite granite in the geode with the size of 1~15mm (photographed at Yansaihu, Shanhaiguan)

2－27　正长斑岩的流动流纹、火山球粒构造(摄于山海关燕塞湖)
Orthophyre with rhyolitic and volcanic spherulitic structure (photographed at Yansaihu, Shanhaiguan)

2－28　火山球粒构造，球粒大小约2~20cm(摄于山海关燕塞湖)
Volcanic spherulitic structure with the spherulite size of 2~20cm (photographed at Yansaihu, Shanhaiguan)

2－29　火山集块岩，紫红、灰绿色，集块成分为安山质，侏罗系髫髻山组（摄于石门寨上庄坨）
Andesitic volcanic agglomerate with the colors of purple-red and greyish-green, Tiaojishan Formation of the Jurassic (photographed at the Shangzhuangtuo Villiage, Shimenzhai Town)

2-30　辉石角闪石安山岩，斑状结构，块状构造（摄于石门寨上庄坨）
Pyroxene hornblende andesite with porphyritic texture and massive structure (photographed at the Shangzhuangtuo Villiage, Shimenzhai Town)

2-31　辉石斜长石安山岩，斑状结构，块状构造（摄于石门寨上庄坨）
Pyroxene plagioclase andesite with porphyritic texture and massive structure (photographed at the Shangzhuangtuo Villiage, Shimenzhai Town)

2-32　角闪石安山岩，斑状结构，块状构造（摄于石门寨上庄坨）
Hornblende andesite with porphyritic texture and massive structure (photographed at the Shangzhuangtuo Villiage, Shimenzhai Town)

2-33　气孔状玄武质安山岩，斑状结构，气孔构造（摄于石门寨上庄坨）
Vesicular basaltic andesite with porphyritic texture and vesicular structure (photographed at the Shangzhuangtuo Villiage, Shimenzhai Town)

2-34　安山质火山集块岩（摄于石门寨上庄坨）
Andesitic volcanic agglomerate (photographed at the Shangzhuangtuo Villiage, Shimenzhai Town)

2-35　火山集块中的烘烤边（摄于石门寨上庄坨）
Baked border developed in volcanic agglomerate (photographed at the Shangzhuangtuo Villiage, Shimenzhai Town)

2－36　火山微球粒旋转构造(摄于山海关燕塞湖)
Rotational structure of volcanic spherulites (photographed at Yansaihu, Shanhaiguan)

2-37 流纹构造(摄于山海关燕塞湖)
Rhyolitic structure (photographed at Yansaihu, Shanhaiguan)

2-38 钾长石旋转斑晶(摄于山海关燕塞湖)
Rotation phenocrysts of K-feldspar (photographed at Yansaihu, Shanhaiguan)

2-39 钾长石旋转碎斑(摄于山海关燕塞湖)
Rotation porphyroclast of K-feldspar (photographed at Yansaihu, Shanhaiguan)

2-40 钾长石球状旋转碎斑(摄于山海关燕塞湖)
Spherical rotation porphyroclast of K-feldspar (photographed at Yansaihu, Shanhaiguan)

2－41　混合岩化片麻岩(摄于北戴河南天门海滨)
Migmatized gneisses (photographed at the Nantianmen seashore of Beidaihe)

2－42　肠状混合岩(摄于北戴河南天门海滨)
Ptygmatite (photographed at the Nantianmen seashore of Beidaihe)

3 地层和古生物

Stratigraphy and Paleontology

实习区地层属于晋冀鲁豫地层区燕辽地层分区秦皇岛小区，为华北型地层。除较普遍缺失上奥陶统至下石炭统、下－中三叠统、白垩系和第三系之外，区内地层出露较为齐全：分别有新元古界青白口系上部、下古生界寒武系和下奥陶统、上古生界上石炭统至二叠系、中生界上三叠统至侏罗系和新生界第四系。

柳江盆地——国家地质公园，位于秦皇岛的北部山区，保存有25亿年以来的太古宙、元古宙、古生代、中生代、新生代各个地质时期的岩石，被地质学界誉为"地质百科全书"。

The practice area belongs to the Qinhuangdao stratigraphic minor region of the Yanliao stratigraphic subregion, Shanxi-Hebei-Shandong-Henan (Jin-Ji-Lu-Yu) stratigraphic region, which represents the typical North China strata. Except the general absence of strata from the Upper Ordovician to the Lower Carboniferous (Mississippian), the Lower-Middle Triassic, the Cretaceous, the Paleogene and the Neogene, the other strata are well outcropped including the upper part of the Qingbaikouan of the Neoproterozoic, and the Cambrian, the Lower Ordovician, the Upper Carboniferous (Pennsylvanian) and the Permian of the Paleozoic, and the upper part of the Triassic and the Jurassic of the Mesozoic and the Quaternary of the Cenozoic.

The Liujiang Basin, as a national geological park situated at the mountain area in the north of Qinghuandao, well preserves the rocks since 2.5Ga relating to the Archean, Proterozoic, Paleozoic, Mesozoic and Cenozoic, which is called "Encyclopedia of Geology" by geologists.

3-1 柳江盆地古生界的地貌景观(摄于驻操营镇东部落村)
The geomorphic landscape of the Paleozoic in the Liujiang Basin (photographed at Dongbuluo Village, Zhucaoying Town)

3－2 新元古界龙山组，石英砂岩夹粉砂岩组成的陡崖（摄于鸡冠山）
Cliff composed of the quartz sandstone interlayered with the siltstone of the Neoproterozoic Longshan Formation（photographed at Jiguanshan）

3－3　新元古界龙山组泥质粉砂岩(摄于鸡冠山)
Muddy siltstone of the Neoproterozoic Longshan Formation (photographed at Jiguanshan)

3－4　新元古界龙山组含铁质砂岩(摄于驻操营镇东部落)
Ferruginous sandstone of the Neoproterozoic Longshan Formation (photographed at Dongbuluo Village, Zhucaoying Town)

3－5　新元古界龙山组不整合于新太古界花岗岩的界线(左)和底砾岩(右)(摄于鸡冠山)
Unconformity contact between the Neoproterozoic Longshan Formation and the Neoarchean granite (left) and the basal conglomerate (right) (photographed at Jiguanshan)

3－6 新元古界龙山组石英砂岩(摄于鸡冠山)
Quartz sandstone of the Neoproterozoic Longshan formation (photographed at Jiguanshan)

3－7 新元古界龙山组石英砂岩中的交错层理(摄于鸡冠山)
Cross bedding developing in the quartz sandstone of the Neoproterozoic Longshan Formation (photographed at Jiguanshan)

3－8　新元古界龙山组石英砂岩中的古波痕(摄于鸡冠山)
Paleo-ripple marks in the quartz sandstone of the Neoproterozoic Longshan Formation (photographed at Jiguanshan)

3－9　新元古界龙山组石英砂岩中的指相矿物——海绿石(摄于鸡冠山)
Glauconite, a facies mineral, in the quartz sandstone of the Neoproterozoic Longshan Formation (photographed at Jiguanshan)

3－10　景儿峪组灰质白云岩(摄于黄土营)
Calcite dolomite of the Jingeryu Formation (photographed at Huangtuying Village)

3－11　景儿峪组灰质白云岩(摄于黄土营)
Calcite dolomite of the Jingeryu Formation (photographed at Huangtuying Village)

3－12　府君山组灰黑色块状沥青质豹皮灰岩(摄于黄土营)
Massive dark grey asphaltene leopard limestone of the Fujunshan Formation (photographed at Huangtuying Village)

3－13 府君山组灰黑色块状沥青质豹皮灰岩(摄于黄土营)
Massive dark grey asphaltene leopard limestone of the Fujunshan Formation (photographed at Huangtuying Village)

3－14 府君山组灰黑色块状沥青质豹皮灰岩(摄于黄土营)
Massive dark grey asphaltene leopard limestone of the Fujunshan Formation (photographed at Huangtuying Village)

3－15　寒武系馒头组，暗紫色泥岩（摄于驻操营镇东部落村）
Dark purple mudstone of the Mantou Formation of the Cambrian（photographed at Dongbuluo Village，Zhucaoying Town）

3－16　寒武系毛庄组、徐庄组，紫色页岩、灰绿色页岩（摄于驻操营镇东部落村）
Purple and greyish-green shale of the Maozhuang and Xuzhuang Formations of the Cambrian (photographed at Dongbuluo Village, Zhucaoying Town)

3－17　寒武系毛庄组、徐庄组，粉砂质页岩(摄于驻操营镇东部落村)
Silty shale of the Maozhuang and Xuzhuang Formations of the Cambrian (photographed at Dongbuluo Village, Zhucaoying Town)

3－18　寒武系毛庄组、徐庄组，灰色钙质泥岩（摄于驻操营镇东部落村）
Grey calcareous mudstone of the Maozhuang and Xuzhuang Formations of the Cambrian (photographed at Dongbuluo Village, Zhucaoying Town)

3－19　寒武系毛庄组、徐庄组，泥质灰岩（摄于驻操营镇东部落村）
Muddy limestone of the Maozhuang and Xuzhuang Formations of the Cambrian (photographed at Dongbuluo Village, Zhucaoying Town)

3－20　寒武系徐庄组,底部的粉砂岩(摄于驻操营镇东部落村)
Siltstone at the bottom of the Xuzhuang Formation of the Cambrian (photographed at Dongbuluo Village, Zhucaoying Town)

3-21 寒武系徐庄组,灰色页岩(摄于驻操营镇东部落村)
Grey shale of the Xuzhuang Formation of the Cambrian
(photographed at Dongbuluo Village, Zhucaoying Town)

3-22 寒武系徐庄组,灰岩(摄于驻操营镇东部落村)
Limestone of the Xuzhuang Formation of the Cambrian
(photographed at Dongbuluo Village, Zhucaoying Town)

3－23　寒武系张夏组，鲕状灰岩(摄于驻操营镇东部落村）
Oolitic limestone of the Zhangxia Formation of the Cambrian (photographed at Dongbuluo Village, Zhucaoying Town)

3－24　寒武系张夏组，竹叶状灰岩(摄于驻操营镇东部落村）
Limestone conglomerates with flat-pebble intraclasts of the Zhangxia Formation of the Cambrian (photographed at Dongbuluo Village, Zhucaoying Town)

3－25　寒武系张夏组，泥质条带灰岩(摄于驻操营镇东部落村）
Limestone with mud-strips of the Zhangxia Formation of the Cambrian (photographed at Dongbuluo Village, Zhucaoying Town)

3－26　寒武系张夏组，灰色灰岩（摄于驻操营镇东部落村）
Grey limestone of the Zhangxia Formation of the Cambrian（photographed at Dongbuluo Village，Zhucaoying Town）

3－27　寒武系崮山组，紫色灰岩（摄于驻操营镇东部落村）
Purple limestone of the Gushan Formation of the Cambrian（photographed at Dongbuluo Village，Zhucaoying Town）

3－28　寒武系崮山组，灰色中厚层灰岩（摄于驻操营镇东部落村）
Grey middle-to thick-bedded limestone of the Gushan Formation of the Cambrian（photographed at Dongbuluo Village，Zhucaoying Town）

3－29　寒武系崮山组，灰色泥质条带灰岩（摄于驻操营镇东部落村）
Grey limestone with mud-strips of the Gushan Formation of the Cambrian（photographed at Dongbuluo Village，Zhucaoying Town）

3－30　寒武系崮山组，紫色砾屑（竹叶状）灰岩（摄于驻操营镇东部落村）
Purple calcirudite flat-pebble intraclasts of the Gushan Formation of the Cambrian（photographed at Dongbuluo Village，Zhucaoying Town）

3－31　奥陶系冶里组，泥晶灰岩（摄于石门寨）
Micrite of the Yeli Formation of the Ordovician（photographed at Shimenzhai Town）

3－32　奥陶系冶里组，豹皮灰岩（摄于石门寨）
Leopard limestone of the Yeli Formation of the Ordovician（photographed at Shimenzhai Town）

3－33　奥陶系亮甲山组，灰岩夹钙质页岩（摄于石门寨亮甲山）
Limestone interlayered with calcareous shale of the Liangjiashan Formation of the Ordovician（photographed at Liangjiashan，Shimenzhai Town）

3－34　奥陶系亮甲山组，灰岩、泥质条带灰岩（摄于石门寨西门外）
Limestone and limestone with mud-strips of the Liangjiashan Formation of the Ordovician (photographed outside the west gate of Shimenzhai Town)

3－35　奥陶系亮甲山组，竹叶状灰岩（摄于石门寨西门外）
Limestone conglomerates with flat-pebble intraclasts of the Liangjiashan Formation of the Ordovician (photographed outside the west gate of Shimenzhai Town)

3－36　奥陶系亮甲山组，风暴角砾灰岩（摄于石门寨西门外）
Breccia limestone by storm processes of the Liangjiashan Formation of the Ordovician (photographed outside the west gate of Shimenzhai Town)

3－37　奥陶系亮甲山组，竹叶状灰岩（摄于石门寨西门外）
Limestone conglomerates with flat-pebble intraclasts of the Liangjiashan Formation of the Ordovician (photographed outside the west gate of Shimenzhai Town)

3－38　奥陶系亮甲山组，风暴角砾灰岩（摄于石门寨西门外）
Breccia limestone by storm processes of the Liangjiashan Formation of the Ordovician (photographed outside the west gate of Shimenzhai Town)

3－39　奥陶系亮甲山组，灰岩中的蛇卷螺化石（摄于石门寨沙锅店）
Ophileta in limestone of the Liangjiashan Formation of the Ordovician (photographed at Shaguodian Village, Shimenzhai Town)

3－40　奥陶系亮甲山组，灰岩中的燧石团块（摄于石门寨沙锅店）
Chert block in limestone of the Liangjiashan Formation of the Ordovician (photographed at Shaguodian Village, Shimenzhai Town)

3－41　奥陶系亮甲山组，泥质团块灰岩（摄于石门寨沙锅店）
Argillaceous lump limestone of the Liangjiashan Formation of the Ordovician (photographed at Shaguodian Village, Shimenzhai Town)

3－42　奥陶系马家沟组(上)与亮甲山组(下)整合接触关系(摄于石门寨沙锅店)
Conformity contact between the Majiagou Formation (upper) and the Liangjiashan Formation (lower) of the Ordovician (photographed at Shaguodian Village, Shimenzhai Town)

3－43　奥陶系马家沟组，白云质灰岩(摄于石门寨)
Dolomitic limestone of the Majiagou Formation of the Ordovician (photographed at Shimenzhai Town)

3－44　奥陶系马家沟组，白云质灰岩中的燧石条带（摄于石门寨沙锅店）
Chert bands in dolomitic limestone of the Majiagou Formation of the Ordovician（photographed at Shaguodian Village，Shimenzhai Town）

3－45　奥陶系马家沟组，白云质灰岩的刀砍纹（摄于石门寨沙锅店）
Differential weathering in dolomitic limestone of the Majiagou Formation of the Ordovician（photographed at Shaguodian Village，Shimenzhai Town）

石炭系
奥陶系

3－46　奥陶系马家沟组白云质灰岩(右)与石炭系本溪组粉砂质页岩(左)之间的平行不整合露头(摄于石门寨)
Parallel unconformity contact between the Ordovician dolomitic limestone of the Majiagou Formation (right) and the Carboniferous silty shale of the Benxi Formation (left) (photographed at Shimenzhai Town)

3－47　石炭系本溪组下部杂色泥岩、页岩、粉砂岩(摄于石门寨)
Variegated mudstone, shale and siltstone of the lower part of the Benxi Formation of the Carboniferous (photographed at Shimenzhai Town)

3－48　石炭系本溪组下部杂色泥岩(摄于石门寨)
Variegated mudstone of the lower part of the Benxi Formation of the Carboniferous (photographed at Shimenzhai Town)

3－49　石炭系本溪组下部杂色砂岩(摄于石门寨)
Variegated sandstone of the lower part of the Benxi Formation of the Carboniferous (photographed at Shimenzhai Town)

3－50　石炭系本溪组底部粉砂质鲕状铝土矿(摄于石门寨)
Silty oolitic bauxite at the bottom of the Benxi Formation of the Carboniferous (photographed at Shimenzhai Town)

3－51　石炭系本溪组页岩(右)与太原组粗砂岩(左)之间的整合接触关系(摄于石门寨)
Conformity contact between shale of the Benxi Formation (right) and the coarse grained sandstone of the Taiyuan Formation (left) of the Carboniferous (photographed at Shimenzhai Town)

3－52　石炭系太原组粗砂岩，球形风化（摄于石门寨）
Coarse grained sandstone with spheroidal weathering of the Taiyuan Formation of the Carboniferous (photographed at Shimenzhai Town)

3－53　石炭系太原组页岩（摄于石门寨）
Shale of the Taiyuan Formation of the Carboniferous (photographed at Shimenzhai Town)

3－54　石炭系太原组砂岩（摄于石门寨）
Sandstone of the Taiyuan Formation of the Carboniferous (photographed at Shimenzhai Town)

3－55　二叠系山西组砂岩(摄于石门寨)
Sandstone of the Shanxi Formation of the Permian (photographed at Shimenzhai Town)

3－56　二叠系山西组碳质页岩(摄于石门寨)
Carbonaceous shale of the Shanxi Formation of the Permian (photographed at Shimenzhai Town)

3-57 二叠系山西组煤层(摄于石门寨)
Coal seam of the Shanxi Formation of the Permian (photographed at Shimenzhai Town)

3-58 二叠系山西组粉砂岩(摄于石门寨)
Siltstone of the Shanxi Formation of the Permian (photographed at Shimenzhai Town)

3－59　二叠系山西组页岩(下)与下石盒子组含砾砂岩之间的接触界线(摄于石门寨)
Contact between shale of the Shanxi Formation (lower) and gravel-bearing sandstone of the Xiashihezi Formation (upper) of the Permian (photographed at Shimenzhai Town)

3－60　二叠系下石盒子组含砾砂岩(左)和砾岩(右)(摄于石门寨)
Gravel-bearing sandstone (left) and conglomerate (right) of the Xiashihezi Formation of the Permian (photographed at Shimenzhai Town)

3-61　二叠系下石盒子组含砾砂岩、砂岩(左)(摄于石门寨)
Gravel-bearing sandstone and sandstone (left) of the Xiashihezi Formation of the Permian (photographed at Shimenzhai Town)

3－62　侏罗系髫髻山组安山质火山岩(摄于石门寨上庄坨)
Andesitic volcanic rocks of the Tiaojishan Formation of the Jurassic (photographed at Shangzhuangtuo Village, Shimenzhai Town)

3－63　侏罗系髫髻山组斜长石安山岩（摄于石门寨上庄坨）

Plagioclase andesite of the Tiaojishan Formation of the Jurassic (photographed at Shangzhuangtuo Village, Shimenzhai Town)

3－64　侏罗系髫髻山组角闪石安山岩（摄于石门寨上庄坨）

Amphibole andesite of the Tiaojishan Formation of the Jurassic (photographed at Shangzhuangtuo Village, Shimenzhai Town)

3－65　侏罗系髫髻山组安山质集块岩（摄于石门寨上庄坨）

Andesitic agglomerate of the Tiaojishan Formation of the Jurassic (photographed at Shangzhuangtuo Village, Shimenzhai Town)

3－66　寒武系生物碎屑灰岩中的叠层石（摄于石门寨）
Stromatolite in bioclastic limestone of the Cambrian（photographed at Shimenzhai Town）

3－67　奥陶系生物碎屑灰岩中蛇卷螺（摄于石门寨沙锅店）
Ophileta in bioclastic limestone of the Ordovician（photographed at Shaguodian Village，Shimenzhai Town）

3－68　奥陶系亮甲山组灰岩中双壳（摄于石门寨沙锅店）
Bivalves in limestone of the Liangjiashan Formation of the Ordovician（photographed at Shaguodian Village，Shimenzhai Town）

3－69　奥陶系亮甲山组灰岩中的腹足类化石（摄于石门寨沙锅店）
Gastropods in limestone of the Liangjiashan Formation of the Ordovician（photographed at Shaguodian Village，Shimenzhai Town）

3－70　奥陶系生物碎屑灰岩(摄于石门寨沙锅店)
Bioclastic limestone of the Ordovician (photographed at Shaguodian Village, Shimenzhai Town)

3－71　奥陶系亮甲山组灰岩中含蛇卷螺化石(摄于石门寨沙锅店)
Limestone of the Liangjiashan Formation of the Orduvician bearing *Ophileta* (photographed at Shaguodian Village, Shimenzhai Town)

3－72　奥陶系亮甲山组灰岩中棘皮类化石(摄于石门寨沙锅店)
Echinoderms in limestone of the Liangjiashan Formation of the Ordovician (photographed at Shaguodian Village, Shimenzhai Town)

3－73　奥陶系海生迹灰岩(摄于石门寨沙锅店)
Limestone with *Thalassinoides* of the Ordovician (photographed at Shaguodian Village, Shimenzhai Town)

3－74　二叠系山西组煤层中的植物化石(摄于石门寨)
Plant fossils in coal seam of the Shanxi Formation of the Permian (photographed at Shimenzhai Town)

3－75　古近系砾石中的植物化石(摄于石门寨马蹄岭垭口)
Plant fossils in gravels of the Paleogene (photographed at the Matiling Pass, Shimenzhai Town)

4 构造地质

Structural Geology

实习区大地构造位置处于中朝地块燕山褶皱造山带的东段，东临太平洋板块。自古生代以来，实习区经历了加里东运动、海西运动、印支运动、燕山运动和喜山运动，形成了一系列近南北向延伸的褶皱与断裂构造。构造运动形迹主要表现有褶皱构造、断裂构造和多级阶地等。

The practice area is tectonically located in the east segment of the Yanshan Fold Orogenic Belt of the Sino-Korea plate east to the Pacific Plate. Since the Paleozoic, a series of folds and faults have been developed in North-South extension by Caledonian Movement, Hercynian Movement, Indosinian Movement, Yanshan Movement and Himalayan Movement. The structural faults are mainly represented by folds, fractures and multistage terraces.

4－1　X型剪节理，发育于新太古界花岗岩（摄于山海关燕塞湖）
X- type shear joints in the Neoarchean granite (photographed at Yansaihu of Shanhaiguan)

4－2　雁行节理，发育于新太古界花岗岩（摄于北戴河老虎石海滨）
En echelon joints in the Neoarchean granite(photographed at the Laohushi seashore of Beidaihe)

4-3 节理，新太古界花岗伟晶岩中发育的“X”型剪节理(摄于北戴河小东山海滨)
X-type shear joints in the Neoarchean granitic pegmatite (photographed at the Xiaodongshan seashore of Beidaihe)

4－4　节理，发育于奥陶系亮甲山组灰岩（摄于石门寨）
Joints in the Ordovician limestone of the Liangjiashan Formation (photographed at Shimenzhai Town)

4－5　节理，发育于新太古界片麻状花岗岩（摄于北戴河海滨）
Joints in the Neoarchean gneissic granite (photographed at Beidaihe seashore)

4－6　节理，发育于新太古界花岗岩（摄于北戴河长寿山）
Joints in the Neoarchean granite (photographed at Changshou Mountain of Beidaihe)

4－7　断层擦痕，发育于新元古界龙山组砂岩（摄于鸡冠山）
Fault striae in the Neoproterozoic sandstone of the Longshan Formation（photographed at Jiguanshan）

4－8　正断层，发育于新元古界龙山组砂岩（摄于鸡冠山）
Normal fault in the Neoproterozoic sandstone of the Longshan Formation（photographed at Jiguanshan）

4－9　由断层错动形成的阶梯状地堑，发育于新元古界龙山组砂岩与新太古界花岗岩（摄于鸡冠山）
Terraced graben developed by faulting in the Neoproterozoic sandstone of the Longshan Formation and the Neoarchean granite （photographed at Jiguanshan）

4－10　逆断层，发育于三叠系砂岩、页岩（摄于石门寨马蹄岭垭口）
Reverse fault in the Triassic sandstone and shale （photographed at the Matiling Pass，Shimenzhai Town）

4－11　褶皱，发育于寒武系灰岩、泥灰岩（摄于石门寨）
Fold in the Cambrian limestone and marlite (photographed at Shimenzhai Town)

4－12　尖棱褶皱，发育于寒武系灰岩夹钙质页岩（摄于石门寨）
Chevron fold in the Cambrian limestone interlayered by calcareous shale (photographed at Shimenzhai Town)

4－13　宽缓褶皱，发育于寒武系灰岩、页岩（摄于石门寨）
Broad and gentle fold in the Cambrian limestone and shale (photographed at Shimenzhai Town)

5 风化作用

Weathering

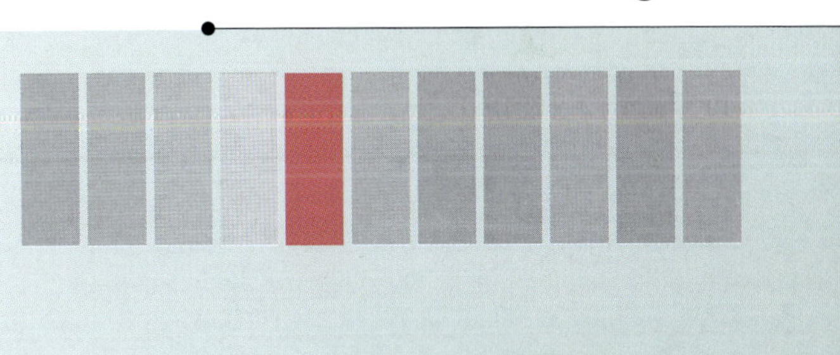

实习区内风化作用地质现象丰富。风化作用是在(近)地表条件下,由于太阳、大气、水、生物等作用使岩石在原地遭破坏的地质作用。风化作用的类型有物理、化学和生物风化作用,其中差异风化和球形风化现象在实习区内非常典型。

There are various geological phenomena originated by weathering in the practice area. Weathering refers to the geological processes that change the physical and chemical characters of rock in situ by sun, air, water, organisms, etc. at or near the surface of the earth. The main types of weathering include mechanical weathering, chemical weathering and biological weathering. Differential weathering and spheroidal weathering are very typical in the practice area.

5-1 风化壳露头,发育于新太古界花岗岩(摄于北戴河山东堡)
Outcrops of weathering crust developed in the Neoarchean granite (photographed at Shandongpu beach of Beidaihe)

5－2　风化壳，发育于奥陶系灰岩(摄于石门寨)
Weathering crust in the Ordovician limestone (photographed at Shimenzhai Town)

5－3　差异风化，发育于花岗岩中伟晶岩脉(摄于北戴河山东堡)
Differential weathering developed in granitic pegmatite vein (photographed at Shandongpu beach of Beidaihe)

5－4　风化壳剖面，发育于新太古界花岗岩(摄于北戴河山东堡)
Weathering crust section developed in the Neoarchean granite (photographed at Shandongpu beach of Beidaihe)

5－5　球形风化，发育于石炭系太原组粗粒杂砂岩(摄于石门寨)
Spheroidal weathering developed in coarse grained greywacke of the Taiyuan Formation of the Carboniferous (photographed at Shimenzhai Town)

5－6　球形风化，发育于细粒花岗岩（摄于山海关燕塞湖）
Spheroidal weathering developed in fine grained granite（photographed at Yansaihu, Shanhaiguan）

5－7　差异风化，发育于太古宇花岗岩（摄于北戴河骆驼石）
Differential weathering developed in the Archean granite（photographed at Luotuoshi，Beidaihe）

6 地下水地质作用

Geological Processes of Groundwater

地下水在运动过程中对周围岩石的溶蚀作用称为地下水的岩溶作用。可溶性岩石在地下水的作用下所形成的独特地形称为岩溶地形。岩溶作用的发育需要满足有可溶性岩石、岩石有透水性、地下水有溶蚀能力和流动等条件。

The erosion of rock caused by groundwater movement is named as the Karst process of groundwater. The distinctive landform constituted by rocks under the erosion of groundwater is called the karst topography. The development of Karst process requires several basic conditions such as the porosity and permeability of rocks, dissolution ability of groundwater, the movement of groundwater and so on.

6－1　岩溶地形(溶沟、石芽等),发育于奥陶系亮甲山组灰岩(摄于石门寨沙锅店)
Karst topography (karren, clint, etc.) developed in the Ordovician limestone of the Liangjiashan Formation (photographed at Shaguodian Village, Shimenzhai Town)

6-2　石芽(摄于石门寨沙锅店)
Clints (photographed at Shaguodian Village, Shimenzhai Town)

6-3　溶沟、石柱(摄于石门寨沙锅店)
Karrens and columns (photographed at Shaguodian Village, Shimenzhai Town)

6-4　落水洞(摄于石门寨沙锅店)
Sinkholes (photographed at Shaguodian Village, Shimenzhai Town)

6-5　溶沟、石芽(摄于石门寨沙锅店)
Karrens and clints (photographed at Shaguodian Village, Shimenzhai Town)

◀▼6－6　溶沟(摄于石门寨沙锅店)
Karrens (photographed at Shaguodian Village, Shimenzhai Town)

7 河流地质作用

Fluvial Geological Processes

河流地质作用是河水及其所挟带的碎屑物在运动过程中对河谷的自然作用，分为河流侵蚀作用、搬运作用和沉积作用。

河流向下冲刷切割河床，使河床不断加深加长的侵蚀作用称为下蚀作用；河水以自身动力及挟带的砂石对河床两侧及谷坡进行破坏的作用称为侧蚀作用。河水在流动过程中，搬运着河流自身侵蚀的和谷坡上崩塌、冲刷下来的物质。当河床的坡度减小，或搬运物质增加，而引起流速变慢时，河流的搬运能力降低，河水挟带的碎屑物便逐渐沉积下来，称为沉积作用。河流沉积作用主要发生在河流入海、入湖和支流入干流处，或在河流的中下游，以及河流的凸岸。

秦皇岛市主要河流有大石河、汤河、新河、戴河和洋河等。

Fluvial geological process is a natural action including erosion, transportation and deposition to the stream valley during the movement of stream and its carried debris.

The erosion can be vertical or lateral to the streambed. The vertical erosion is resulted in the downward scour and cutting of the stream to the streambed and can deepen the stream valley. The destruction to the two banks of the streambed and valley walls is called the lateral erosion as the stream swings from side to side across its valley floor carrying sand and gravels. During stream flowing, the sediment particles eroded by stream and collapsing from valley walls will also be transported by the running water. When the slope of streambed reduces or the transported sediments increase, the stream's velocity will be lowered. Then the stream transportation capacity will decrease so that the sediments carried by running water will be deposited. The stream deposition is commonly developed at or near the end of a stream, in the middle and lower reaches of a stream, or at the convex bank.

The main streams in Qinghuangdao are Dashi River, Tang River, Xin River, Dai River and Yang River etc.

7-1 “V”形谷(摄于石门寨义院口)
V-shaped stream valley (photographed at Yiyuankou Village, Shimenzhai Town)

7-2 河曲(摄于石门寨上庄坨)
Meandering stream (photographed at Shangzhuangtuo Village, Shimenzhai Town)

7-3 河床与阶地(摄于石门寨上庄坨)
Streambed and stream terraces (photographed at Shangzhuangtuo Village, Shimenzhai Town)

7－4　河漫滩、边滩沉积物(摄于石门寨上庄坨)
Overbank and point bar deposits（photographed at Shangzhuangtuo Village，Shimenzhai Town）

7－5　浅滩(滨河床浅滩)沉积物(摄于石门寨上庄坨)
Deposits near streambed (photographed at Shangzhuangtuo Village, Shimenzhai Town)

▲▼7－6　新河河口三角洲(摄于北戴河新河河口)
Estuary delta of the Xin River (photographed at the estuary of the Xin River, Beidaihe)

7－7　心滩(摄于北戴河新河河口)
Midchannel bar of the Xin River (photographed at the estuary of the Xin River, Beidaihe)

7－8　沙嘴、沙坝(摄于山海关大石河河口)
Spit and sandbank（photographed at the estuary of the Dashi River，Shanhaiguan）

7－9　潟湖(摄于山海关大石河河口)
Lagoon（photographed at the estuary of the Dashi River，Shanhaiguan）

8 风的地质作用

Aeolian Geological Processes

风的地质作用主要发生于植被稀少、地表物质疏松、蒸发量远大于降雨量的干旱和半干旱地区，也常见于滨海地区，风的侵蚀、搬运、沉积作用以机械方式进行。

The aeolian geological process occurs mainly in the arid and sub-arid areas with rare vegetation, loosing surface and more evaporation than rainfall and sometimes also in the seaside. The erosion, transportation and deposition of wind act in mechanical way.

8－1　风成沙丘(摄于秦皇岛昌黎县翡翠岛)
Aeolian dune (photographed at Feicuidao, Changli County, Qinhuangdao)

8-2　风成沙丘(摄于秦皇岛昌黎县翡翠岛)
Aeolian dune (photographed at Feicuidao, Changli County, Qinhuangdao)

8－3　风蚀穴(摄于秦皇岛昌黎县)
Blowouts (photographed at Changli County, Qinhuangdao)

8－4　风成交错层理(摄于秦皇岛昌黎县)
Aeolian cross bedding (photographed at Changli County, Qinhuangdao)

8－5　风成沙波纹(摄于秦皇岛昌黎县)
Aeolian ripple marks (photographed at Changli County, Qinhuangdao)

8－6　风成沙波纹(摄于秦皇岛昌黎县)
Aeolian ripple marks (photographed at Changli County, Qinhuangdao)

9 海洋地质作用

Marine Geological Processes

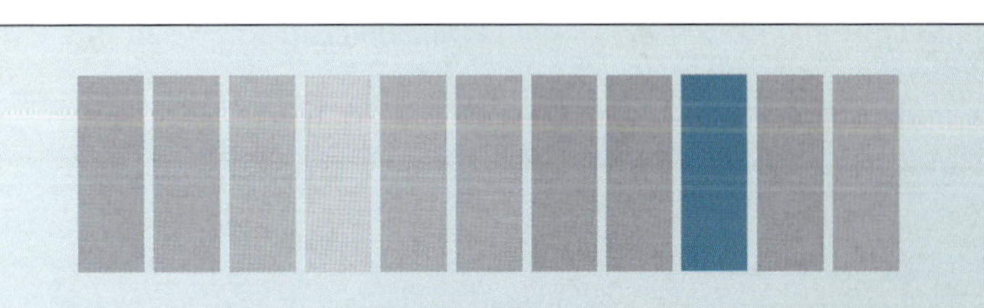

北戴河实习区位于渤海湾沿岸,海洋地质现象十分丰富,对海洋地质现象的观察和认识是大学生在北戴河实习的重要内容。

海洋地质作用包括海蚀作用、搬运作用和沉积作用。在波浪冲蚀基岩海岸时,最先在贴水处形成海蚀凹槽。凹槽扩大,上部崩塌,形成海蚀崖。随着陡崖后退,海蚀凹槽的底部逐渐扩大为向海微微倾斜的基岩平台,称为海蚀平台(波切台)。海平面下降或陆地上升,海蚀平台明显高于海平面而呈现的阶梯状地形,称为海蚀阶地。海蚀平台在波浪作用下,坡度渐缓;当海浪对海蚀平台破坏力趋于零,这时的海蚀平台横剖面称为基岩海岸平衡剖面。由于构成海岸的岩石及构造的差异,抗蚀能力不同,海蚀作用形成海蚀洞、海蚀穴、海蚀桥、海蚀柱等地形。

Beidaihe practice area is located at the Bohai seaside with various phenomena of marine geology. Thus the observation and cognization to the marine geology is an important and special content for the undergraduates in the practice program.

Marine Geological processes are reflected by marine abrasion, transportation and deposition. The wave erosion to the rocky coasts will produce wave-cut netch near the water surface and then sea cliffs along with the collapse of rocks above the caves at the headland. In pace with the cliff retreat towards to the land, abrasion platform (wave-cut platform) will be formed and slightly tilted towards to the sea. When the sea level drops or the land is uplifted relatively, formed abrasion platform will be obviously higher than the sea level and is called as abrasion terrace. Under the wave action, the slope gradient of abrasion platform will gradually decrease until the wave destruction to the abrasion platform becomes zero. Then the cross section of the abrasion platform reaches a balance and is named as the equilibrium profile of the rocky coast. Due to the variety of lithology and structure of the rocks, marine abrasion can create different landforms such as sea caves, sea arches, sea stacks and so on.

9-1 拍岸浪(摄于北戴河老虎石海滨)
Surf (photographed at Laohushi seashore, Beidaihe)

9－2　海蚀崖(摄于北戴河鸽子窝海滨)
Sea cliff (photographed at Geziwo seashore, Beidaihe)

9－3　海蚀沟(摄于北戴河老虎石海滨)
Sea trenches (photographed at Laohushi seashore, Beidaihe)

9－4　海蚀凹槽(摄于北戴河南天门海滨)
Wave-cut notch（photographed at Nantianmen seashore，Beidaihe）

9－5　海蚀坑(摄于北戴河小东山海滨)
Sea cave（photographed at Xiaodongshan seashore，Beidaihe）

9－6　海蚀礁(摄于北戴河小东山海滨)
Sea stacks（photographed at Xiaodongshan seashore，Beidaihe）

9－7　海蚀穹(摄于北戴河南天门海滨)
Sea arches（photographed at Nantianmen seashore，Beidaihe）

9－8　海蚀沟(摄于北戴河小东山海滨)
Sea trenches（photographed at Xiaodongshan seashore，Beidaihe）

9－9　海蚀崖(摄于北戴河鸽子窝海滨)
Sea cliff (photographed at Geziwo seashore, Beidaihe)

9－10　海蚀岩垛(摄于北戴河鸽子窝海滨)
Sea stacks (photographed at Geziwo seashore, Beidaihe)

9－11　海蚀礁(摄于北戴河老虎石海滨)
Sea stacks (photographed at Laohushi seashore, Beidaihe)

9－12　波切台、海蚀礁(摄于北戴河小东山海滨)
Wave-cut platform and sea stacks (photographed at Xiaodongshan seashore, Beid-aihe)

9－13　古海蚀凹槽(摄于北戴河小东山海滨)
Paleo-wave-cut notch (photographed at Xiaodongshan seashore, Beidaihe)

9－14　古海蚀穴(摄于北戴河老虎石海滨)
Paleo-sea caves (photographed at Laohushi seashore, Beidaihe)

9－15　古海蚀沟(摄于北戴河老虎石海滨)
Paleo-sea trenches (photographed at Laohushi seashore, Beidaihe)

9－16　古海蚀凹槽(摄于北戴河老虎石海滨)
Paleo-wave-cut notch (photographed at Laohushi seashore, Beidaihe)

老虎石、小东山沿岸可见一级和二级阶地沙质海岸地质现象。由松散沙粒组成的海岸，地形较为平坦，在波浪和潮汐作用下，进流和退流带动沙粒垂直海岸方向运动，造成沙粒的反复运动。

The first and second terraces with geological phenomena of sandy coast can be observed at Laohushi and Xiaodongshan seashore. The relatively flat coast is constituted by loose sand grains. Under the action of wave and tide, current of landward and seaward will carry sand grains to move vertically to the coast alternately.

9－17 退潮后出露的水下沙坝（摄于北戴河山东堡海滨）
Underwater sand bar during tide ebbing (photographed at Shandongpu seashore, Beidaihe)

9－18 涨潮（摄于北戴河山东堡海滨）
Flood tide (photographed at Shandongpu seashore, Beidaihe)

9－19 高潮线（摄于北戴河山东堡海滨）
High tide line (photographed at Shandongpu seashore, Beidaihe)

9－20　沿岸沙滩(摄于北戴河山东堡海滨)
Longshore beach (photographed at Shandongpu seashore, Beidaihe)

9－21　波痕、海白菜(摄于北戴河山东堡海滨)
Ripple marks and sea lettuces (photographed at Shandongpu seashore, Beidaihe)

9－22　层理(摄于北戴河山东堡海滨)
Beddings (photographed at Shandongpu seashore, Beidaihe)

9－23　潮道(近景)(摄于北戴河鸽子窝海滨)
Tidal channels (close shot) (photographed at Geziwo seashore, Beidaihe)

9－24　心滩(摄于北戴河山东堡海滨)
Barrier island (photographed at Shandongpu seashore, Beidaihe)

9－25　水下沙坝(摄于北戴河鸽子窝海滨)
Underwater sand bar (photographed at Geziwo seashore, Beidaihe)

9－26　潮道(远景)(摄于北戴河鸽子窝海滨)
Tidal channels (long shot) (photographed at Geziwo seashore, Beidaihe)

9－27　障碍流痕(摄于北戴河鸽子窝海滨)
Current mark across barrier (photographed at Geziwo seashore, Beidaihe)

▼►9－28　连岛沙坝(摄于北戴河老虎石海滨)
Tombolo (photographed at Laohushi seashore, Beidaihe)

波痕是滨海、浅海环境中常见的沉积构造类型，是沉积环境分析的重要标志。

波痕由波峰、波谷和波长等要素组成，一般成组出现。通常按波痕形态分为对称波痕、不对称波痕、平顶波痕、槽状波痕等。

Ripple mark. It is among the common types of sedimentary structure and thus an important index for analyzing the marine sedimentary environment.

A ripple mark can be described by factors of crest, trough and wavelength and often appears in groups. Usually, according to the shape, ripple mark can be presented as symmetric ripple mark, asymmetric ripple mark, flat topped ripple mark and so on.

9－29　干涉潮流（摄于北戴河新河口）
Interference tide (photographed at the estuary of the Xin River, Beidaihe)

9－30　干涉波痕(摄于北戴河新河河口)
Interference ripple mark (photographed at the estuary of the Xin River, Beidaihe)

9－31 脊状波痕(摄于北戴河山东堡海滨)
Ridge-like ripple marks (photographed at Shandongpu seashore, Beidaihe)

9－32 槽状波痕(摄于北戴河新河口)
Trough-like ripple marks (photographed at the estuary of the Xin River, Beidaihe)

9－33 不对称波痕(摄于北戴河新河口)
Asymmetric ripple marks (photographed at the estuary of the Xin River, Beidaihe)

9－34　平顶波痕(摄于北戴河新河口)
Flat topped ripple marks (photographed at the estuary of the Xin River, Beidaihe)

9－35　纵向波痕(摄于北戴河山东堡海滨)
Longitudinal ripple mark (photographed at Shandongpu seashore, Beidaihe)

生物遗迹。生物体在沉积物表面或内部活动的证据，包括运动、栖息、掘穴、觅食等行为的痕迹。

Biological relics. It means evidence of the activity of organisms preserved on or in sediments and includes the relics of moving, resting, burrowing and feeding.

9－36 生物遗迹（摄于秦皇岛市南戴河昌黎海滨）
Biological relics (photographed at Nandai River－Changli seashore)

9－37　生物遗迹(摄于北戴河海滨)
Biological relics (photographed at Beidaihe seashore)

9－38　气泡沙(摄于北戴河海滨)
Sand with "bubbles" (photographed at Beidaihe seashore)

滨海生物。北戴河海滨生物种类丰富，具有垂直分布的特征。

Littoral organisms. At Beidaihe seashore, there are abundant and diversified organisms with vertical distribution.

9－39 贝壳滩(摄于北戴河小东山海滨)
Shell beach (photographed at Xiaodongshan seashore, Beidaihe)

9－40 基岩海岸海洋生物分带，短滨螺，牡蛎、褐藻、海白菜(摄于北戴河小东山海滨)
Marine organism zones at rocky coast with *Littorina* sp., oyster, brown algae and sea lettuce (photographed at Xiaodongshan seashore, Beidaihe)

9－41　藤壶(摄于北戴河小东山海滨)
Barnacles (photographed at Xiaodongshan seashore, Beidaihe)

9－42　牡蛎、短滨螺、海白菜(摄于北戴河小东山海滨)
Oyster, *Littorina* sp. and sea lettuce (photographed at Xiaodongshan seashore, Beidaihe)

9－43　紫贻贝、海白菜(摄于北戴河小东山海滨)
Mytilus edulis and sea lettuce (photographed at Xiaodongshan seashore, Beidaihe)

9－44　黑偏顶蛤、海白菜(摄于北戴河小东山海滨)
Modiolus atrata and sea lettuce (photographed at Xiaodongshan seashore, Beidaihe)

9－45 短滨螺(摄于北戴河小东山海滨)
Littorina sp.（photographed at Xiaodongshan seashore，Beidaihe）

9－46 牡蛎(摄于北戴河小东山海滨)
Oysters（photographed at Xiaodongshan seashore，Beidaihe）

9－47 锈凹螺、寄居蟹(摄于北戴河小东山海滨)
Chlorostoma rustica and Paguridae（photographed at Xiaodongshan seashore，Beidaihe）

9－48　海蟑螂(摄于北戴河小东山海滨)
Ligia sp.（photographed at Xiaodongshan seashore，Beidaihe）

9－49　笠贝(摄于北戴河小东山海滨)
Notoacmea（photographed at Xiaodongshan seashore，Beidaihe）

9－50　樱蛤(摄于北戴河小东山海滨)
Tellinidae（photographed at Xiaodongshan seashore，Beidaihe）

9－51　鹿角菜
Pelvetia sp.

9－52　褐藻
Brown algae

9－53　海葵(开放)
Sea anemones (open)

9－54　海葵(闭合)
Sea anemones (close)

9－55　褐藻
Brown algae

9－56　绿藻
Green algae

9－57　海蚯蚓
Arenicola sp.

9－58　生物遗迹
Biological relics

9－59　水母
Jellyfish

9－60　海星、螃蟹
Starfish and crab

9－61　章鱼
Octopus

9－62　海星
Starfish

9－64　扁玉螺(猫眼)
Neverita didyma（cat's eye）

9－65　毛蚶
Clam（*Scapharca* sp.）

9－63　贝壳滩
Shell beach

9－66　蛤蜊
Clam（*Mactra* sp.）

9－67　斑尾复鰕虎鱼(虎斑钱)
Spottedtail goby (Tigroid money / Hubanqian in Chinese)

10 矿产资源与环境保护

Mineral Resources and Environmental Protection

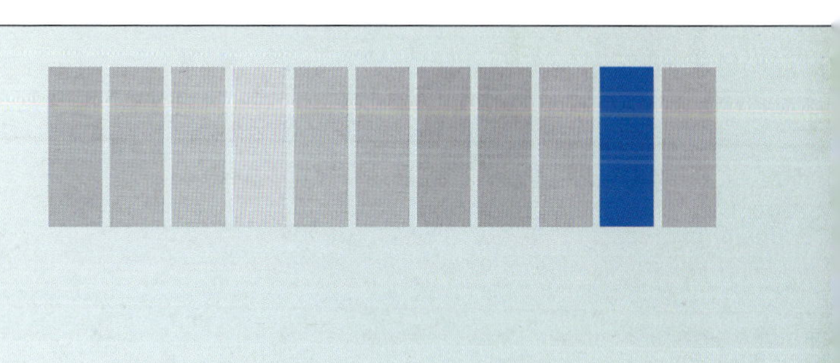

实习区矿产资源丰富，主要有煤、铝土矿和耐火黏土、石灰岩、石英砂岩、花岗岩等。

In the practice area, there are abundant mineral resources including coal, bauxite, fire clay, limestone, quartz sandstone and granite etc.

10-1 水泥厂（摄于石门寨）
Cement factory（photographed at Shimenzhai Town）

10-2 石灰岩采石场（摄于石门寨）
Limestone quarry（photographed at Shimenzhai Town）

10-3 鲕状铝土矿（摄于石门寨）
Oolitic bauxite（photographed at Shimenzhai Town）

10－4　煤矿（摄于石门寨）
Coal mine（photographed at Shimenzhai Town）

环境保护。合理利用自然资源，保持人与自然和谐发展。

Environment protection. Reasonable utilization of natural resources is a way to maintain the harmonious development between human and nature.

10－5　采矿造成的地裂缝（摄于石门寨）
Ground fissure caused by mining (photographed at Shimenzhai Town)

10－6　采矿造成的地面塌陷（摄于石门寨）
Collapse caused by mining (photographed at Shimenzhai Town)

11 旅游地质资源

Geological Tourism Resources

实习区所在的秦皇岛市自然人文旅游资源极为丰富，北戴河、南戴河、“山海关”万里长城等是中国旅游胜地和国家重点文物保护单位。

Qinhuangdao, where the practice area is located, has plentiful natural and cultural tourism resources. Beidaihe, Nandaihe and Shanhaiguan Pass of the Great Wall etc. are China Tourist Resorts and National Cultural Heritage Sites.

11－1　山海关
Shanhaiguan

11－2　山海关——“天下第一关”门楼
Shanhaiguan — The gatehouse of “The first Pass of the world”

11－3　山海关城墙
City wall of Shanhaiguan

11－4　万里长城起点——老龙头
The start of the Great Wall — Old Dragon Head（Laolongtou）

11－5　澄海楼
Chenghailou

11－6　宁海城
Ninghaicheng

11－7　御碑亭
Imperial Stele Pavilion（Yubeiting）

11－8　海神庙、天后宫(摄于老龙头)
Queen of Heaven (Tianhou) Palace and Poseidon (Haishen) Temple (photographed at Laolongtou)

11－9　角山长城(摄于角山)
The Great Wall on Jiaoshan (photographed at Jiaoshan Mountain)

11－10　长城遗址(摄于义院口)
The Great Wall Ruins (photographed at Yiyuankou)

11－11　孟姜女庙(摄于孟姜女庙)
Temple of Lady Meng Jiangnü (photographed at Temple of Lady Meng Jiangnü)

11－12　鸽子窝海湾日出
Sunrise at Geiziwo Gulf

11－13　鹰嘴石
Eagle mouth Stone（Yingzuishi）

11－14　北戴河鸽子窝公园(摄于鸽子窝)
Geziwo Park in Beidaihe (photographed at Geziwo Park)

11－15　北戴河鸽子窝海湾
Geziwo Gulf of Beidaihe

11－16　老虎石公园
Laohushi Park

11－17　犀牛望月(摄于老虎石公园)
Rhinoceros looking up at the moon (Xi niu wang yue) (photographed at Laohushi Park)

11－18　惊涛拍浪(摄于老虎石公园)
Surging wave breaking against the shore (Jing tao pai lang) (photographed at Laohushi Park)

11－19　北戴河山东堡海湾（摄于山东堡）
Shandongpu Gulf of Beidaihe（photographed at Shandongpu seashore）

11－20　秦皇岛市石门寨城墙遗址
City Wall Relics of Shimenzhai Town，Qinhuangdao City

11－21　秦皇岛市石门寨“点将台”遗址
Commander Appointment Stage Relics of Shimenzhai Town，Qinhuangdao City

11－22　秦皇岛市长寿山公园
Changshoushan Park of Qinhuangdao City

11－23　骆驼石（摄于北戴河骆驼石）
Camel Stone（photographed at Luotuoshi of Beidaihe）